ELIMINATING MA BELL

ELIMINATING MA BELL

Table Of Contents

(Click On Any Chapter To Be Taken There Immediately.)

Introduction

The one definitive thing that should be said about VoIP technology is that it will transform your life. There are many ways in which it can do this, too. You've heard commercials and seen a million different advertisements for VoIP. Now, it's time to learn all that you can about this technology so that you can add it to your life.

VoIP is one type of technology that offers mostly benefits and it is definitely not hard to use. You'll be ready to go quickly and you'll be able to start saving money right away, one of the largest reasons to switch your long distance service to VoIP.

Throughout this e-book, we'll explore what VoIP is and how it will transform your life. You'll be able to explore the options that are available to you, learn how to use VoIP in your home and in your business and you'll quickly see how you'll save a small fortune on it.

Being fair, there are a few disadvantages to this technology, but we'll show you how to completely take advantage of all that VoIP can offer even around these consideration. If you are on the edge trying to decide if you want to switch to VoIP, we'll help you to make that decision.

The largest benefit that VoIP promises is that it will provide you with high quality long distance services without the cost that usually goes along with it. What the long distance companies don't want you to know is that it is completely simple to use and it will allow you to call locations around the world, easily.

Instead of going with just what the VoIP sales people are offering or counting on what the long distance carriers are saying, take the time to fully learn what the options are that you have.

Learning about VoIP for yourself will allow you to make the best decision, regardless of the sales talk that you will find everywhere.

Here is a view of VoIP that is honest and fair and that will provide you with the information you need to make a decision about this service.

Read through this book and make your own decision about its worthiness in your home and your business.

Chapter 1: Just What IS This VOIP Stuff Anyway?

What have you heard about VoIP? Now, forget about it. What you should learn about it you'll learn here. Remember, sales marketing has really done a job on VoIP to this degree.

What Does VoIP Do?

When you install VoIP technology into your home, you'll have the ability to talk to others using Voice Over Internet Protocol (VoIP.) All that means is that you'll be able to use the phone that's connected to the internet to make calls.

Instead of phone lines that traditionally run over the electrical or phone lines in your home already, VoIP uses your internet connection instead. Your standard phone will plug into the VoIP adapter to be used.

In most cases, you won't be able to tell the difference in using VoIP over other phone technology. It gives you the same basic results. It sounds easy, right? It really is that simple.

VoIP Technology

With the help of VoIP technology, you can communicate over your internet lines, which will also happen to be a no costing way to do so.

To do this, several key things will be necessary.

1. You will need to connect your standard telephone to an adapter unit that is VoIP qualified.

2. Or, you may instead use a microphone equipped computer to make your calls instead of a phone.

3. When you pick up the phone to dial your friend across the ocean, the VoIP adapter realizes what you are doing and turns the voice signals you are sending into your phone into digital based signals. These are then sent across the ocean through the Internet, instead of traveling through any phone line.

4. Your friend receives your call in the same way. He or she will pick up the phone when you call and then VoIP goes to work again. This time it translates the digital like signals back into a standard voice transmission. When your friend uses their phone, they hear your voice, nothing more.

As you can see, it really doesn't make much of a difference in the experience that you have using the telephone. You'll be doing the same thing that you are already doing.

Here's how it works:

Phone Call –To- Adapter- Internet - Adapter –To- Phone

Although it may seem like this process takes time, it really does not. It will take little to no time to convert you voice into a digital signal to go over the internet service that you have and back into your voice so your friend will hear you. In fact, the process happens so fast you may not even realize what is happening at all and not realize you are using it!

Computer Communication

Another option that you may have comes in the form of using the computer in your home to make your long distance calls. For example, if you would like to call your friend, instead of using your telephone to make the call, you could use your computer.

In order to do this, you will need to have a computer that has a microphone located in it.   This microphone will serve as your communication tool.

When you move to make a call, you will simply use the software that is installed on your computer to call your friend. When they answer, you can talk to them without any type of handset. Instead, you'll use the microphone to talk into.

The process is still the same. The computer will take your voice and translate it into a digital signal which is then sent over the internet to your friend's computer. They can either use the phone or their own computer to talk to you. The voice once again changes from a digital signal into an actual voice and you hear your friend, just as you would if you had dialed their phone number into your phone.

You can also do this by using a computer that is equipped with a headset that allows you to speak into it for more privacy.

As you can see, VoIP technology is quite easy to use. You'll be able to install it into your computer and phone system quickly and easily. You'll be able to get it in and using it quickly.

In a business environment, VoIP works about the same way. In most cases, though, it is even better because it call allow any size of business to stay connected to their clients, suppliers and business associates around the world.

If you are considering VoIP technology for your business, you'll find additional information about that in later chapters. What you should know now is that it is an ideal tool for communication in any field and in any use. It will help you to stay in connection as well as let you to save on costly long distance.

The benefits of using VoIP are easy to spot. It is easy and it saves you money! But, beyond this, there are additional benefits in having the ability to connect to others around the world. If you are ready to pick up your phone and order it, don't do it just yet. You need to understand a bit more about how it works and why it works. You also need to understand what you need to make it work in your personal home or in your business environment.

Chapter 2: Here's What You Gotta Have For VOIP

In order for you to take advantage of VoIP, you will need to have some equipment. The good news is that it is not complex, difficult to install and it's not generally that expensive.

The one requirement that you simply can't go without is broadband or high speed internet connections.

What Is Broadband?

Understanding what broadband internet service is happens to be important. If you don't have this type of connection yet to the internet, you may want to use it. If you don't' have it, you are probably using dial up access to connect to the web.

While there isn't anything wrong with dial up, the way that broadband works makes connecting to the internet faster and even simpler. When you use this method of connection, you'll be able to get online faster and even surf through the pages you want to go to quicker. That allows many benefits in and of itself.

Broadband is quite fast and it is the way of the future in technology for internet access. It may be a bit confusing to understand how broadband technology actually works, as it is somewhat complex, it is easy to understand just how beneficial it can be!

Broadband works by allowing for a larger amount of data to move faster all at one time. A good way to understand it is to consider your water pipes.

If you replace your old pipe (dial up internet) with a larger, wider pipe (broadband) you are now allowing much more information to get through and therefore allowing the water to move through it easier.

The larger that the pipe is, the more water can go through it. The same is true with broadband. With the higher bandwidth that it uses, broadband can get you around the internet faster and even gets you off the telephone line you are currently using with dial up access.

The benefits of broadband include the ability to download information faster, the ability to connect to websites, to view movies and yes, even to talk to other people using your internet connection. That's where VoIP comes in.

VoIP technology simply could not work well over dial up access to the internet because it is not fast enough and it can not carry the signals fast enough.

In order for you to use VoIP, you need to have broadband internet access. You can use this type of internet access through your current cable modem, through your DSL or even through a local high speed cable network offer.

No matter which of these you have, VoIP technology is likely to be able to be used with it.

Your Adaptation

Before you can get the VoIP technology, you'll also have to have the right tools to communicate over the internet. As we mentioned earlier, you can use your computer with a microphone or a headset with a microphone as well as an adapter for your telephone. They also make specialized phones for VoIP too.

The style of these services that you need or can use will depend on the type of VoIP that you select. Since VoIP is not a brand name, so to speak, there are actually several services that you can purchase it from. That means that there are differences in each of the carriers that you use VoIP through.

Some VoIP technologies work through the computer only. For this type of application, you'll need to be able to use a microphone or a headset to communication. You may be able to use this service with the help of a specialized phone as well.

Many other services allow you to use a standard phone but with the help of an adapter. Here, you'll need to purchase the adapter that will plug into your phone and into your modem.

If you are using a computer to use VoIP services, you will need the right software installed into your computer. Generally, this comes to you when you purchase VoIP services. The microphone that you need can be quite inexpensive, generally under $20.

If you go with the specialized phone to connect to VoIP services, you'll be able to purchase a phone that will plug right into the broadband connection that you have. It doesn't need to plug into the phone jack or your computer in any way.

The specialized phones work just like a traditional phone in that you'll pick up the receiver, dial the number and instantly get the person on the other line to talk to.

Finally, if you use VoIP through a standard telephone with an adapter, you will be able to dial the same as you do right now.

The only difference is that your phone may not have a dial tone (in most cases.) Your service provider generally provides this type of service to its clients and home users easily.

You should know a bit more about the adapter that you may be using. It is called an ATA or Analog Telephone Adapter. In order to provide service to you, this device connects to your broadband internet service connection using a router, the third component in using VoIP.

To do this, it will use an Ethernet (RJ45) cable and it will connect to your phone using a telephone cable (RJ11) to complete the circuit.

The ATA is necessary for transferring your analog information into the usable digital media that can be carried over the internet so that you can chat with others. The adapter must then send this digital signal to the router of you computer which then will determine the right route to take to get to the right network and finally to the person that you were planning on calling. This happens instantly, though

Your Router

Many people that have broadband internet service already have a router in place. That's because it is necessary for you to if you have more than one computer that you wish to connect to the internet. If not, you'll need to purchase one. They do not usually come with your VoIP package, but can for an additional charge.

The router's job is to connect multiple networks together. Its job is to connect your home's network system with the internet, which technically is the world's largest networking system.

If you only have one computer that's connected to your service to the internet or any type of device connected to the internet, then you may not need a router at all. You could simply attach your VoIP service right to the broadband service that you have. Yet, this is simply uncommon today.

Those that do have more than one device that will connect to the internet, then a router is a must. A router will help you to connect to the internet at each point that you have the need. Routers enable you to share one connection to the internet with several devices at the same time. This allows you to log into your internet connection from more than one location.

In order to do this, your router will use Network Address Translation, which is also called NAT. This technology allows you to share one IP address which is assigned to your computer by your internet service provider as a distinction of who you are.

One benefit that a router can provide to you is the ability to provide a firewall to your network. A firewall can protect your computer users from all sorts of things like spyware, malware and viruses from getting in. A firewall can help to keep problems like this from infecting your computer and ultimately bringing down your computer network.

Even if you do only have one computer, or one device that requires internet connectivity, you still need a firewall to protect your internet access.

Those that have the need for a router will also find it to provide many other useful features in regards to internet usage. For example, you can block specific websites from being traveled to, which is wonderful advantage for a home with children.

Or, you can determine a time to cut off usage of the internet in the home. This is helpful to those that want to turn off their connection (even their VoIP service) at any time.

Even better, routers are excellent tools for the wireless technology you have. You can use routers to help you to connect your PDA, your MP3 player, or your laptop (any device that wants and can get online for that matter) to the internet. It can do this without have to deal with wires because of the wireless technology which it provides to your home.

Summing Up This Chapter

In other words, you will need:

- Broadband internet access
- Either: A standard phone with a VoIP adapter

        Or, a computer that features a microphone.
        Or, a specialized VoIP phone

- Router

That's it! There are no costly wiring projects to worry about and you won't need to do crazy configurations with your internet service either (or your computer for that manner.)

In the way of tools for VoIP, there is nothing else you need. Later, you'll learn some tools that you may want to consider adding to your service as well.

Once you have subscribed to the VoIP service, they will help you to decide the right type of technology for their service and for your needs.

It is also important to point out that much of what you will need to get VoIP up and running in your home will come to you with the help of your VoIP provider. Most providers will present you with the VoIP adapters that you need for your phone. You'll also be able to get the necessary software for your computer.

There are some tools that you need to get or have on your own, though. For example, your headset or microphone for using VoIP over the computer is up to you to purchase. That goes for your router too. These things are not all that expensive, though. A router may cost you under $50 depending on where you live and your needs for it.

Your broadband service is something you also need to subscribe to which does offer an additional cost per month to use. But, depending on the type of service that you have right now, you may actually save money by switching from dial up internet to broadband internet access. Look into companies that offer internet access to compare prices and speed.

Chapter 3: How Do I Make A Call?

As we have mentioned, using VoIP is easy to do. When you are looking to make a call to someone next door or someone on the other side of the planet, the process is basically the same thing.

There are some differences that range widely from one carrier of VoIP types of service and the type of product that you have to make calls. Nevertheless, it is an easy to understand process.

Local And Long Distance

The most common question that is asked of individuals that are considering switching to VoIP is that of how it works for local calls as well as long distance calls.

This changes somewhat from one vendor to the next, but in most cases, VoIP doesn't charge a per rate call on customers that use their VoIP service to contact other individuals that have VoIP. In most cases, the provider of the VoIp service will allow you to use this service to call others that have it.

Some providers of VoIP will allow you to make calls to individuals that live in another area code other than where you live.

In addition, you should realize that although you may be not charged for making long distance calls, those that call you that do not have VoIP will still incur the charges of a long distance call, based on what their long distance provider requires.

Some VoIP services are different. For example, some providers may charge you much like your current long distance plan charges, just at a large discount. This may be that you are charged a long distance call when you call individuals that live outside of your calling area. This would be set up quite like the standard, wire line telephone service currently is.

Additionally, VoIP providers are sometimes offering their calling services at a flat rate, again, like wire line telephone services. You may have a selected number of minutes to use per month, too.

Before you sign up with any type of VoIP service provider, you should find out what they have to offer as well as the costs for each option they offer so that you can compare.

Who Can You Call With VoIP?

Another aspect of using VoIP is who you can call. Each provider offers a different definition of this and for that reason you should look into this more thoroughly to allow you to compare the options that you have.

Here is a look at some of the options out there.

Limited: Some VoIP providers only allow you to call other people using VoIP that already have VoIP themselves. In this case, if you were to call someone that doesn't have VoIP, you would have to use a standard long distance plan or you wouldn't be able to connect to them. If you call someone that has VoIP, you would connect to them through that service.

Unlimited: Some VoIP providers will allow you to call anyone that has a phone number. This may not be limited to home phones, either. You may be able to call anyone that has a phone number even those that have mobile phone/cell phone numbers, long distance numbers and even international phone numbers.

With unlimited use, you'll be able to call others that do not have VoIP. Here, they would not need to have any specialized equipment to talk to you. They could use their standard home phone.

Providers also range in the services that they provide with VoIP calling. For example, some may allow you to talk to several people with VoIP service at one time, while others do not offer this service.

The features and benefits of each provider of VoIP is something you'll learn more about later. We'll teach you how to choose the right provider of your service!

Here are a few other questions that are often asked about VoIP:

Does my computer need to be turned on to use it?
No! There is no need to worry about if your computer is on or off. It will work for both needs. If you need to make a call using your computer, then it will need to be one. If your VoIP runs through your computer, it needs to be on.

Can I use my computer and be on the phone at the same time?
Yes, in most cases, you can be on both the computer and your VoIP service.

Can I take VoIP with me when I travel?
Believe it or not, many of the best providers of VoIP service are offering these services. It depends greatly on your provider.

How do I know if there's a phone call coming in?

If you have a traditional phone that uses an adapter for VoIP, your phone will ring just as it would if you were using a standard, analog phone. If you are using your computer, your computer will alert you when there is a call coming in using VoIP through the software that's installed.

Does it sound good?

While many worry that it will sound choppy or it will be too quiet to hear, that's not the case. VoIP works well for those that use it and you won't be able to tell the difference.

What about wireless computers and hot spots?

If you travel with your laptop by your side, and you have VoIP service through that laptop, chances are good that you'll be able to use your VoIP service at hot spots located in coffee houses, in diners and even in airports. Your provider must offer this service, though.

Now, let's move on to more topics about VoIP.

Chapter 4: Here Are The Pros and Cons Of VoIP

There are both good and bad things about VoIP. Most of the time, it is an excellent tool for accomplishing the tasks that you want and need to do. In fact, it may just be the phone of the future in the next decade, but for that to happen, the need for high speed internet access must be met by all that desire it.

As you are considering VoIP, you must pay attention to the needs that you have in it. There is much to consider both with advantages and disadvantages.

The Advantages

It is always good to start with the positives, and VoIP technology does offer many of them. This is especially true if you already have your service set up in the way of already having a broadband connection and having the necessary equipment (microphone, headset, and/or router.) If you have this, then you are already set to be benefited by all that can be offered to you.

No Land Line: One of the first benefits that you get by having VoIP is the need to remove your land line telephone service. Don't worry, there's no need to dig up the yard here. Simply disconnecting from expensive long distance is one of the best benefits that VoIP will offer to you. Without the need for a traditional landline, you are able to stop paying for it, too!

Unlimited Communication: Some of the better VoIP service providers allow you to communicate with people around the world for as long as you want, with little cost. For example, if you were to purchase a flat rate service for VoIP, you would be able to connect to your friend in Japan that has it and talk to them over the internet for literally hours on end without the fear of cost.

This is a benefit that people have been seeking simply because of how devastating it can be to not pick up the phone and call a loved one because it is too expensive to do so. Now, you can easily do so without the worry of cost with a flat rate service.

Multiple Connections: Most VoIP service providers allow you to communicate to others as you see fit. But, what if you want to talk to several people all at one time?   This is called 3way calling on traditional land lines and it can be a costly investment which is why most people don't even bother to get it added to their phone lines.

With VoIP, you can connect and talk to many people, as many as you want, without having to pay this additional fee. That can leave you with just the benefits that you are looking for in VoIP.

No Broadband Cost: With some of the better VoIP service providers, you'll be able to stop paying for your broadband service because it will be part of your VoIP connection. This means that you can lower your costs of connecting to the internet significantly just by using VoIP technology.

You will need to find out if you can actually get this benefit by asking your VoIP service provider if your broadband service is included in your VoIP package.

Low Cost: Even with the most expensive VoIP service providers, you still get the benefits of having a low cost when you use VoIP technology over long distance technology. The costs are lower because there are less people and companies involved in your transaction.

Since your connection is traveling through the internet, it doesn't need to go through landlines, which continuously need to be repaired, maintained and new lines need to be put in. Instead, the internet's big network allows you to connect to it without much cost, other than the cost of the broadband service you have.

The cost to make a traditional call from your computer to a traditional land line is still quite inexpensive in the realm of traditional long distance calls.

No Cost: In some cases, there is no cost to using VoIP technology. You will be able to connect to the internet with your broadband technology and talk to someone else that has VoIP using your computer without cost. Computer to computer conversations may be at no cost to you, by some VoIP service providers.

Let's say that you live in California and you want to call your good friend in France. When you do so over the internet using your VoIP technology that you both must have and make this connection over the use of your computer, you don't have to pay for the interaction. In this regard, your conversation, even overseas, is free.

Take It With You: When you use VoIP as you travel, you literally can pick it up and take your connection, and your low long distance rates, to a minimum. That's because most VoIP service providers provide you with the ability to make VoIP portable with you.

For example, if you are traveling, you may have had to pick up the phone and call your spouse with high, very high, costs. Have you noticed the in room charges for long distance like this?

But, when you use VoIP technology, this doesn't happen. You can use your same VoIP plan that you are using to connect to others while you are traveling. To do this, you do need to have a broadband connection. That isn't as much of a problem as you may think with coffee houses and airports (even shopping malls for that matter) offering "hot spots" or internet connectivity.

To make this happen, you will need to use your headset or you will need to have a VoIP specialized phone that allows you to do this, which may come as part of your service package. Just sign into your service and then you can instantly make your call.

When you use VoIP to travel with, you don't have to worry about your cell phone minutes or even have to worry about the roaming charges you may be prone to if you use your mobile phone. There's no need to worry about cell phone coverage or long distance fees when you use VoIP in this manner.

Portable With You: Almost the same as using VoIP while you travel is the benefit of being able to use VoIP for all of your portability needs. You can use VoIP technology to allow you to make phone to phone VoIP calls from virtually anywhere that there is a connection to the broadband internet.

For example, if you are in Rome and you want to call someone either next door or around the world, you can do so using your VoIP service, even if it is based out of your home in the United States. Your VoIP service provider will give you a VoIP phone number that will follow you around the world, wherever you decide to go.

By doing this, you can make calls anywhere, no matter where you are, using your VoIP service. All you would have to do is to plug your VoIP service into any broadband service provider. Of course, it would have to be a VoIP capable phone that you are using.

Light Phones: Along with this traveling and portability that VoIP can provide to you, we should mention that the portable phones that you will use to make this happen are just like cell phones. They are light, easy to use and they allow you to make VoIP calls, which may even be free. What's not to love about them?

Free Features: Another benefit to using VoIP is that you get all of your favorite features in using traditional calling without cost.

This includes things like call waiting, caller identification, call forwarding, three way calls, and voicemail and so on. You can use these services without any type of problem too. It is part of the ease that VoIP makes making calls.

Many traditional phone lines make you pay for these services either as an additional service or as a cost per use feature. In either case, the costs can be high and are not necessarily all that worthwhile with a traditional phone line. But, with VoIP, it's free.

Send Data: With VoIP, you can also communicate through data without a cost to it. For example, you can send data from one business to the next business, without an additional cost in doing so. You can send photos to your family and friends, send messages to your family, using text services, and even download and send pictures too. There is quite a lot of benefit to doing this as the costs of doing so otherwise, such as with a fax or having to use postage even, are much higher.

As you can see, there are many different benefits to using VoIP technology. You'll easily be able to plug into this service and get the most out of it. If you are interested in any of these things, like being able to talk free, to travel and talk without roaming charges or even just to be able to send pictures without having to pay an arm and a leg for them, then VoIP is definitely something you should consider.

The Disadvantages

Being fair to you, it is important to provide you with a look at some of the problems that come from using VoIP technology. While many of these problems were first brought out with great concern, there have been new advances that can help you to overcome some of the worry that you have over them.

Powers Needed: One of the largest problems that face those that are switching to VoIP technology is the need for electricity. While you home is currently up and running with it, you don't think about it. But, what happens when the power goes out?

Should the power go out on you when you have VoIP service, you may lose your connection to the internet which means that you may lose your ability to pick up the phone and all others. In a time of crisis, this is an important service that you need to have.

Some VoIP service providers are allowing you to have or are even providing themselves, some type of power backup. If the power were to cut off, you would have the ability to use the backup if it was necessary to do so. But, not all providers offer this ability.

In order to overcome this problem, you do need to invest in a backup or you need to insure that your VoIP provider is offering it to you.

No White Pages: Another of the disadvantages of VoIP stem from the inability of being able to use white page directory. For example, because you are not using a traditional landline, your telephone number may not be able to be used in a white page listing.

You also may not have access to white page listings when you need to look up someone's phone number, such as when you are calling 411 for information. On the other hand, you should be able to do this when you are just searching for information from your computer terminal anyway.

This is also something that is changing with more and more companies that offer VoIP offering white page directory service to you.

Find out what your VoIP service provider has to offer you in this regard.

VoIP And 911

One of the largest concerns regarding VoIP is that it may not connect to local 911 services. This is actually one of the largest and most important considerations that you have. But, in recent months, new technology has helped to bridge the gap at least somewhat.

Here's the problem. When you pick up the phone to call 911, you need help. If you can not access 911 for your public safety and emergency preparedness needs, you can be placed in trouble and the FCC has been working on resolving this problem for some time.

When VoIP technology first came onto the market, consumers began contacting the FCC in the United States in an effort to complain about not being able to dial 911 and get their local police, fire or rescue services fast enough to their home. This inability to contact emergency services has left people in a public safety problem that's critical in regards to the potential risks.

The Federal Communication Commission has studied and has come back with new requirements for VoIP technology. It has imposed the new Enhanced 911, which is also called E911.

With older technology for VoIP, you could have picked up the phone and dialed 911, but you may or may not have been able to get in touch with the right agencies. But, even if you did, there was no way for those governmental protection agencies to help you unless you could give your physical address.

Because the 911 technology of old did allow for this ability to tell the emergency services where you were and even gave them your call back number, they could get to you faster, saving lives.

With Enhanced 911, this is installed again. The FCC has instituted laws that require that those that have interconnected VoIP services (which is what you would be using) will need to use the Public Switched Telephone Network (called PSTN) to provide your information. This network must include wireless networks, too.

This means that if you are to originate or terminate a call, it must be used on this network in order to insure that individuals can still gain access to emergency services. The Enhanced 911 makes sure of this by automatically providing your call back number as well as your location information, in most cases, to emergency personnel.

Understanding Interconnected VoIP Service And Public Safety

Interconnected service is a term that the government uses to describe VoIP. When it was first made available, VoIP did not take into consideration such public service problems, but today this has changed.

With a standard telephone service, the phone number that you have is associated with your address. This means that your phone and your location are locked in, making it easy for emergency workers to know where you are located in case of emergency.

But, with VoIP, this is not necessary possible in that regard. VoIP services allow you to take your phone with you, no matter where you go. While this is a huge benefit for the VoIP user, it still presents a problem in that you could be any place that there is a broadband connection.

That means that there is not one set location for your phone number to be connected with your location.

Within the last few months, VoIP technology has been adjusted, so to speak, to insure that the person that dials 911 can be able to take advantage of public safety services. Even still, there are things you need to know as the consumer using VoIP.

Traditional Phone: You pick up the phone to call 911 because you hear someone breaking into your home. As soon as your call is picked up, the emergency worker on the other end already knows your location and the phone number you are calling from. This is because the Public Safety Answering Point (the location in which your call is received which is generally assigned to one particular area) has the ability to track your call.

With VoIP: Calling your emergency personnel using VoIP is different. Unlike what was described above, VoIP service has some challenges that you, the consumer needs to realize.

Some of the problems that users can anticipate and may have a problem with, according to the FCC include these problems when you dial 911:

1. Your VoIP service may not connect to the PSAP (your local emergency dispatch system.)

2. Your VoIP service may call the administration line of your PSAP instead of calling the emergency connection number. This number may not be staffed with personnel to aid you. Or, trained 911 dispatchers may not be available.

3. You may get through to the emergency service provider, but it may not provide them with your location information or your phone number information.

4. You may need to provide your location and other vital information to the emergency provider if you are using VoIP services. If your location changes, you'll need to update them on this change, too. This is the information they'll need to find you.

5. VoIP may not work during a power outage. If there is no internet connection, or that internet connection fails for some reason (even if it gets overloaded while you are using it) you lose your ability to connect to VoIP services and therefore to the emergency workers.

In an effort to stop these problems, there have been a number of updates and requirements put in place for VoIP service providers.

1. 911 services must be provided as part of the standard service of VoIP. It becomes a mandatory feature, not something that you may not have or have to request to have. You can't opt out of 911 coverage either.

2. Before you can get your VoIP service, the FCC has made the requirement of providing your physical address of where the primary service will be used. This will help emergency teams to find your primary location. VoIP providers must also insure some easy way that you can report your current location if it should change from this location.

3. Your VoIP service provider must provide information about where you are. All 911 calls must be sent to the right location with emergency service providers in your local area. This includes all callback numbers and your location.

4. If there are any limitations on you using 911 with your VoIP service, the provider must provide this information to you before you sign up for the service. That means presenting all risk that you have for not being able to connect effectively to 911 for any reason must be disclosed. You must also sign off saying that you understand the risks of using VoIP service as your only means of communicating with emergency services in your local area.

Understanding How This Affects You

What does all of this mean to you, the subscriber to VoIP who may one day need to call on emergency services? There are actually many things that you'll need to realize and do if you are to keep yourself safe and still take advantage of what VoIP can offer to you.

First and foremost, you should make sure to provide your physical address to your provider of your service, so that they can communicate this when it becomes necessary. Keep your address updated too. If you leave and take VoIP with you for a weekend trip, report this to your VoIP provider.

You should also understand just what your limitations are regarding VoIP. You need to realize that others using your phone also need this information. This includes the babysitter and your kids, too.

You'll need to install and maintain a back up power supply so that your VoIP service can run when there is a power failure. Find out if your VoIP provider offers any type of benefit like this. If you have a mobile/cell phone, this can act as a back up, but only if you keep it fully charged so that it can be available in an emergency.

With an understanding of some of the limits that VoIP presents to you, you can actually get an understanding of what you need to do in order to get the advantages of VoIP without becoming overwhelmed with the various problems that can arise.

For those that are considering VoIP, understanding the advantages and disadvantages of VoIP is essential. If you are unsure of what it can provide you, or you are concerned about one of the disadvantages, contact your VoIP provide to find out just what they can offer you in the way of information.

Remember too that VoIP technology is different from each provider of it. You'll need to see how each of these providers has interpreted the FCC rulings on VoIP and have installed safeguards for them. The good news is that most providers have made sure to provide you with the information and tools to keep you safe and protected even in the worst of times.

Check out what the providers offer as safeguards to 911 and other problems. See what additional benefits they may be able to offer to you, too!

Chapter 5: Can I Use This In My Business?

Although most of the marketing of VoIP is directed at the in home consumer, one of the largest benefits that this service can offer is to the business owner.

Just consider your business phone bill. Do you pay hundreds if not thousands of dollars each month to keep your employees connected with clients, shippers, suppliers and the rest of their needs? Do you pay long distance charges that are exceedingly high? Are you afraid to even look at the international charges on your phone bill?

VoIP is designed to allow both the personal and the advanced business owner to have all of the interconnectivity that they need at the affordable rates that only VoIP can offer to you.

If you are a business owner, VoIP will cut your costs down considerably. It can even provide additional services to you to enhance your experience with VoIP. Here, we'll show you only some of the tools and benefits of using VoIP with your business.

Costs Drastically Lowered

The largest benefit to the business owner in using VoIP is the cost savings. There is no doubt that the long distance costs that most business interact with on a daily basis are going to be outrageous with a traditional phone line, especially when business associates and customers are located around the world.

With VoIP, your calls are transmitted into data which is then sent over the internet. It works just in the same way that it would in a home user environment. With this also comes the benefit to having the low costs of making calls and sending information.

Here's an example. Let's say that you are on your way to a meeting and running late. You get a business call on your laptop telling you about an important update. You take it then head out to the airport. When you get there, you get an urgent voicemail that was directed to your business phone. The good thing is that it was just sent to your email inbox and now that you've got a hotspot, you've got the information.

Although you are traveling from New York to Los Angeles, you didn't have to have your customers, your clients or even your business associates change the place they call you. You take your phone number (with New York area code) with you to Los Angeles and don't miss a single beat in the process.

Because all of your information and data is sent over the internet, the costs are low. Business accounts are readily available to the business owner for VoIP services. You won't have to worry about getting the most for your money, especially with flat rate service. Just compare for one minute the difference in cost from a standard or traditional long distance plan to what you can get with VoIP. You'll be amazed.

Another money saver that the business has when they use VoIP is that they don't have to maintain separate networks for their data transmissions and their phone system.

You don't have to incur additional costs for when an employee moves, changes or is added. This can easily cost a $100 or more per employee when it happens on a traditional line. This way, it costs you nothing. You just take the phone and move it to where they are going and plug it in.

Productivity

Just imagine all that you can do and keep on doing because there is the ability to stay connected no matter where you go. There are features that allow you to do all of the things that you would want to do and then so much more along with it.

One of the largest benefits that come from using VoIP for increased productivity has to do with the way that your phone numbers are configured.

For example, you can configure the system so that a call that comes in will be simultaneously ringing on several devices. For example, it could go to the desk or to the cell. That means that before it heads to your voice mail account, it's been able to be answered faster and by more locations. Therefore, you don't miss important calls. You stop the problem of phone tag!

This service along was shown by one research firm, Sage Research, to have increased productivity in the workplace by adding almost 4 hours per week per employee to your role. That is a huge significance to many business owners today worrying about productivity.

There are additional ways that it can benefit you and your employee productivity, too.

For example:

- Plan a business meeting where you and your associated are located around the world. The good news is that they are right there in the same room with you using video phone technology. The cost is the same as a phone call using your VoIP service.

- In the middle of the night you can get your messages without leaving your home. No need to worry about the piece of information that you forgot to bring home. Its available right on your VoIP service.

- Need to send a message to your employee? Don't bother with picking up the phone to call them, having to do all the formalities or even waiting for them to become available. Just send them a message that arrives right in their email box.

- You don't have to wait for a line to be available. As long as you have a VoIP ready phone or a computer with a microphone, you can talk to them in seconds.

- Access your business phone anywhere, anytime as long as you have a connection to the broadband service. In airports, at home, or at the hotel around the world that you are staying with.

- Talk with your laptop. With the right software installed on your laptop, you can send and receive all of your calls using a microphone or headset. Take your laptop with you and gain the benefits.

- Voicemail and faxes can easily be sent and received through VoIP. They will end up in your email inbox. That means that you have on easy place to get them, and you can easily forward them to others.

- Use virtual phone numbers to connect with your customers. If you want your customers to believe that you are in one location and that they are dialing a local company, select a phone number with that locations area code.

Getting VoIP Into Your Business

So, we've convinced you that there are benefits to adding VoIP into your business, but how do you do that?

The business systems that you have in place are all complex and while VoIP really is simple for you to install and start using, even in the business world, there are things to consider before installing it.

One thing you should consider right away is hiring a professional to handle the process of installation of the service. Even using a VoIP service provider's expert installation professionals is a smart move to make for a number of reasons.

But, this isn't necessary if you have a small business and are using a standard VoIP product. You'll be able to install this type of system without a problem in most cases. If you do have more dedicated plans, though, you should consider hiring a professional.

A larger VoIP system is just as beneficial and can even provide more to you than a standard service. But, hiring a professional is almost always necessary just because of the details. You can also have your service tailored to fit your specific needs when you have a professional install them. It's not that costly and it will pay off within the first few months of use.

Testing, Testing...

You should also implement your VoIP technology slowly at first. By taking the time to test one area of your business with VoIP will work out and glitches or concerns that people may have. By testing VoIP on just a few users will help you to make sure that you are happy with using VoIP and that you have the system set up to provide you with the tools that you need.

By testing the VoIP service with just a handful of employees, you can insure that the process is going to be beneficial to you. The good news is that in most cases, working out the bugs takes minutes if that and you'll be able to start benefiting from it right away.

Call Forwarding Benefits

Another thing to take into consideration is call forwarding services. For example, the largest problem with VoIP is that if the power goes out or you somehow otherwise end up with no internet service (as if that's every happened!) you'll need some method of still being able to use your phone system.

There are two solutions you should install to prevent this problem. First, consider having a back up power source available. Depending on your desired need, having a generator (the size that would equip your business with what it needs) will help you to actually keep your internet up and running even when the power goes out. Installing a generator or other back up system doesn't have to be too costly or too time consuming either.

Secondly, you should consider using a call forwarding set up. That means that for when the power goes up, you have your VoIP system set up to move calls to a landline phone (the traditional type) or they can be forwarded to a cell phone or even a service.

By instilling these things into your system, you'll save yourself from not having the ability to communicate when you need to.

Security And VoIP

One thing that you must do when it comes to using VoIP is securing your network. If you will be using VoIP for any important considerations, or your internet connection is important to you and your business, then you need security. This should not be something that is new to you in that there are countless products and reasons to use security if you have an internet connection at all.

But, with VoIP you want to insure that not one can get into your business through your internet connection. Technically, without the right firewall and protection on your system, a hacker could get in and wreak havoc on your business phone service.

But, before you throw VoIP out the window because of this risk, realize that with the right protection, you can completely safeguard your internet connections.

In fact, you are no more at risk with the VoIP technology as you are with the internet you are using.

As a business owner, it is important to consider all of the tools that are available to your business to make it run as smoothly as possible. Without taking advantage of something that potentially could save you thousands of dollars in profits every month, you are simply wasting your money.

VoIP is good for business. It cuts costs. It increases your employee's productivity. It helps your business to stay connected. It provides you with high quality, above standard quality phone service. It just works for business needs.

When shopping for a VoIP provider is a major concern. Later in the e-book you'll be able to understand the options that you have as well as what to look for. Shopping for a provider is very easy to do when you have all of the benefits of doing so on the web. Of course, you'll want to compare several companies to find out which VoIP provider offers you the very best option for your business needs.

Chapter 6: How Do I Choose?

You have quite a bit of information at hand, now. You probably have realized that the benefits of using VoIP incorporate many things for the home user as well as for the business owner. Now, you need to find the right VoIP provider.

Not all providers are the same, as you can tell from other areas of this e-book. It pays a great deal to look at what each provider has to offer to you and then decide which one is the ultimate right choice for your needs. The good news is that you are likely to find many benefits available to you by comparing the providers offering service in your area.

In this chapter, we will look at the things that are essential for you to implement into your provider choice. Luckily, with this information you can compare what is available by providers easily and make a decision based on it.

Costs

The costs of VoIP service varies from one provider to the next more so than anything else. You can save up to 80 percent of your phone bills using VoIP, in most cases.

Shopping around will help you to find out who has the lowest rates and the plan that fits you the most. For example, you may want to consider an unlimited service plan, which allows you to make calls as often as you want, when you want to. This can start at under $20 per month but extended up to nearly $60.

Quality, features and the number of customers your provider has determines the cost of the same service options that you have. Remember to compare apples to apples in that you should compare your VoIP unlimited options to your local calling plans, etc.

Features

The second largest consideration that you have in service provider options is that of features. As we have mentioned, you'll find that the features that a VoIP provider offers are excellent in comparison to a traditional phone service. But, not all providers offer the same ones or even come close to delivery the same number of features.

Depending on what you need and want in features will depend on which provider you select. Again, though, while these features may have cost you a per use or a monthly charge, they are included into the price here.

Your Phone Number

For many, keeping the same phone number is a must. Just like changing your cell phone number or keeping it the same when you change companies, the same happens with VoIP providers that offer it. Some providers will offer this benefit of allowing you to keep your number while others will not.

You shouldn't want to switch your phone number right off the bat when switching to VoIP. This is especially true if you are not familiar with the service and want to give it a test run first. Nevertheless, it should be a consideration that you have when it comes to choosing a VoIP service provider.

Find out which service providers will actually allow you to keep your old number and which will not.

Calling Internationally

Another important consideration that you have is using your VoIP calling internationally. Depending on your need for this, both business and personal use, you should take into consideration your needs.

If you make a lot of calls to international numbers, then you need to actually invest the time in finding out what international rates are for the locations that you call.

Some VoIP providers offer different rates for international calls than they do for local calls. In addition, there may be some restrictions on where you can call. There are some carriers, few, though, that offer unlimited calling plans to certain countries, which can be an ideal benefit for those that are looking for a way to stay connected.

Still, by far the rates of international calling using VoIP are far lower than the costs of using a traditional phone calling plan.

911 Security

You'll find a lot of information about how 911 works with VoIP in earlier sections of this e-book. But, what you need to know about choosing a VoIP service provider is important. Most providers do offer Enhanced 911, but not all do. You need to make sure that this is something that is offered or that you have a back up plan in place for.

Trial Runs And Warranties

Another thing for you to take the time in considering is that of guarantees and trial offers. Most of the VoIP providers are offering some type of trail run to insure that you have the necessary time to determine if VoIP is right for you. If you take the time to insure this information, you'll be well on your way to testing out the service and what it can offer to you.

You should also look for the money back guarantee. Check to find out what the provider offers in the way of a money back guarantee. You want a product that is going to provide you with quality service, but since VoIP is very new, having this type of guarantee helps considerably.

Choosing a VoIP provider isn't difficult since most offer their services right on the web. But, you should invest the time in searching for several providers until you find one that fills all of your needs accurately.

Chapter 7: Hooking It All Up

There are some very simple steps that are required when you go to put VoIP into your home. You have a couple of different options as far as installation as well equipment that you can use to complete your installation.

Different Options On Accessing The VoIP

- You can hardwire your VoIP system to your home through the use of a VoIP adapter. This is as simple as plugging the adapter into your existing phone jack and allowing the other lines to feed through it.
- You can purchase a multi handset cordless phone system which will allow all of the phones to have the access to the VoIP system since they will all be connected through one phone jack.
- The last option is a wireless jack which uses the electrical wires in your home to transmit the VoIP signal to all of the phones in your home.

Any of these options will work when trying to get your initial signal to your phone or phones while using the VoIP system. When considering any of these systems you need to take into consideration what your needs are as well as how much that you are looking to spend on your system.

Pros And Cons Of Each Installation Method

- Hardwiring is your most reliable option as it will provide a central location for the access of the system from all of your points. It is also the most economical as it does not require the purchase of any new phones for your home. This system also has low noise and interference levels.
- A multi handset cordless system is the easiest of the three methods as it requires simply plugging your base station into the VoIP receiver. You need to make sure however that the phone signals will not interfere with your wireless receiver.
- Wireless receivers are the least reliable of the three methods as they rely on your electrical lines to act s the phone lines for your system. Each telephone must be connected to its own wireless receiver as well as the base.

Steps For Connecting To Your VoIP Network

1. Purchase a surge suppressor to insure reliability.
2. Connect your VoIP adapter to your broadband provider; make sure to verify how this should be done with your individual provider.
3. Next, you should disconnect your old phone wiring system from your existing phone lines.
4. Finally connect and distribute your VoIP system throughout your home.

Make sure that you check with your individual provider for the specific installation instructions as each provider has their specific stipulations and requirements as far as how to install and distribute your system.

Access To The VoIP Network

Your access to the VoIP Network is brought into your home through your broadband connection. In most cases this can be brought into your home through your existing phone line or through your cable provider. In order to use the VoIP technology you must have broadband access to your home. This can be provided by either your local or you may even be able to access it through your VoIP provider if they are multifaceted.

Chapter 8: Some Advanced Options

There are many options that you now have available to you through the VoIP system. Some of these benefits include call recording, teleconferencing and even portability.

Call Recording

With many VoIP systems there is now the ability to record calls that you have placed to someone. This can be used with your business or if you would like to listen to a call you are having with someone that you want to save. This is not however an anonymous item; meaning that the person on the other line will be made aware that the call will be recorded. This is done when you click the record button on your system a message will come over the other parties handset that will tell them that recording has begun. When you are done recording a message will play telling the other party that the recording has ended.

This can be very beneficial in terms of business as it will allow you to record dealings or negotiations as well as to be able to go back and correct any misinformation that you may have entered if you are entering information for shipping or for legal reasons.

Teleconferencing

This is a very useful feature that is available through your VoIP system. It is very easy to access as you only need two small items to be able to use this. The two items that you will need are a webcam and a microphone. This allows you to use your computer as a video telephone which helps create a more personal atmosphere as creating better interaction between the parties involved in the conversation.

This simple process takes advantage of the equipment you already have and allows you to use it to make phone calls with your VoIP system over your broadband connection to your computer. The VoIP video phone feature can come in very handy in regards to long distance relationships as well as allowing different parties to se each other and better understand through pictures and demonstrations right through your phone.

Portability

When you add the VoIP service to your business you are creating a new benefit for you and your employees. Stability, at least as far as phones are concerned has been created.

With the VoIP system you are able to transfer phone lines from one office or one station to another without the need to reroute or wire. It is very simple. You just take your handset with you. The handset that you are using will have your designated number assigned to it. You are then able to go and take your phone with and plug it into the VoIP jack at your new station or office.

This little feature will not only create continuity in your office, but with your client base as well. This would mean the end to misrouted calls as well as getting a hold of the wrong person because their office moved.

In all these features can be very beneficial to just about any business and in some cases even your home. If you are willing to take advantage of them and learn how to use them properly to enhance your profitability as opposed to having technology, but not knowing how to use it. Use them to the fullest and see what these features can bring to you.

Investigate the service you will use fully to see which features they offer before settling on one.   This will allow you to take advantage of what may be offered instead of missing out completely.

Wrapping It All Up

You've had the opportunity to consider all that VoIP can offer to you. You'll find constant additions to the products and services that VoIP can offer to you as this product continues to grow.

Many believe that VoIP will take over the calling abilities of the world, making it easier and significantly less expensive to call others. But, until high speed internet connectivity is available readily around the world, that's not going to happen. Yet.

The good news is that there are many benefits to VoIP at this point. You'll be able to install it within a few minutes and start benefiting from it right away.

If you want options in your phone use and you realize the benefits that VoIP can provide, there's nothing to stop you from really getting what you want.

www.ingramcontent.com/pod-product-compliance
Lightning Source LLC
LaVergne TN
LVHW021305200726
843509LV00012B/1795